FLORA OF TROPICAL EAST AFRICA

CYATHEACEAE

PETER J. EDWARDS

Erect ferns with stems (trunk or caudex) to 10 m in Africa; rarely (not in Africa) prostrate or epiphytic; vascular tissue forming a hollow cylinder perforated by large gaps, the surface of stems covered with densely matted adventitious roots, rarely spiny; stipe bases persistent or, if not, then stipe scars prominent, old fronds rarely clothing the stem. Fronds almost invariably clustered at top of stem in a very tight, rosette-like spiral. Stipe bases completely covered with scales, these often on wart-like or spiny epidermal outgrowths, lateral pneumatophores (aerophores) forming discontinuous lines on stipe and rachis; upper (adaxial) surface of stipe grooved, nearly always bearing many short, multicellular, antrorse hairs. Lamina herbaceous to coriaceous, 1–3-pinnate, very rarely entire (not in Africa), often oblong, elliptic or deltoid in outline, truncate at base or with one or more lower pinnae gradually or sharply reduced, rarely with very reduced to skeletonized pinnae ('aphlebiae') at the stipe base; rachis with upper surface usually with antrorse hairs, as on stipe. Pinnae-rachis (costae of 1-pinnate species) grooved, usually hairy (less often also with scales) on upper surface, this axis and the lesser axes of costule (midrib of pinnule) and veins variously scaly on the under (abaxial) surface, the scales progressively smaller towards the apices of each subsequent division, very small, 2–3-celled appressed hairs present on the smaller axes, including veins. Veins simple or forked, usually free, rarely fused to form areoles (not in Africa). Sori hemispherical, with or without indusia; indusium either attached all around receptacle base and often covering the young sorus, opening to form a firm-edged cup or shallow disc (in some species breaking irregularly), or attached only on the costular side of the receptacle ("hemiteloid" condition) and often of very variable size, the larger examples appearing to be full cups or discs before loss of sporangia reveals their obvious asymmetry; exindusiate species may have small narrow scales around the base of the receptacle. Sporangia obovoid, with angularities (due to packing), developing in basipetal succession, borne on a conspicuous ± spherical or columnar receptacle to 1 mm high, annulus complete, oblique, stalk very short, 4-rowed (4-seriate), often mixed with multicellular hairs with or without a glandular apical cell. Spores trilete, usually 64, sometimes 32 or 16 per sporangium, perispore smooth, pitted, longitudinally or radially ridged or folded, verrucate to echinate, also rodlets. Chromosome number n = 69.

A large family considered here as including a single genus, *Cyathea*. Tryon in 1970 proposed a new classification with six genera; this system was widely followed in America. Several recent publications on African ferns have adopted these genera and place the African species in the segregate genus *Alsophila* R. Brown. As noted by Kramer the classification is not consistent if macromorphological and spore characters are considered. Readers are referred to Holttum (1981) for further details on the species of *Cyathea* in Africa. This account is based on his work and the specimens he examined at Kew between 1978 and 1981. His research has been reinterpreted by examination of much new material and of his annotated material using better microscopes with higher resolution and better lighting than were available to him.

1

Note: **stipe and rachis colour** is not stated except for *C. mildbraedii* and *C. fadenii* where they are consistently dark brown throughout, a diagnostic feature of Holttum's Group 1 (Holttum 1981). Although colour may be important in living material it is very variable in herbarium specimens, even in the same collections or even fronds. Different degrees of exposure, even drying techniques, may contribute to this variation. It appears that younger/smaller axes of most species are paler in colour.

Scales: the larger scales on the stipe and rachis of all indigenous species are attached to hard outgrowths which remain as small warts ("verrucae") or spines (in *C. manniana, C. mildbraedii*). In *C. cooperi* epidermal scars remain after the scale is shed. Smaller scales of the stipe, rachis and pinna-rachis are often common, but are not mentioned unless diagnostic. Incidentally, many of these are also attached to tiny warts and it is these warts which produce the variously rough texture found in, at least, the stipe-bases of all species. The scales (and hairs if present) on the lower surface of the pinnules (costae, costules and veins) are often diagnostic in form, distribution and density and are therefore discussed for each species near the end of the descriptions. In all species they are glossy.

Hairs: the groove on the upper surface of the rachis has scales present or absent but almost always has very persistent, short (usually under 1.5 mm), characteristic stiff antrorse, sharp pointed, colourless, transparent or, more usually, opaque or sub-opaque, whitish or pale brown multicellular hairs. These hairs are usually glossy and do not collapse noticeably, if at all, on drying. Similar but spreading hairs, with or without a flaccid apex, may also be present. Lesser axes have variously expressed levels of smaller scales and/or hairs, usually with stiff antrorse hairs present on the upper surfaces of some axes. Lower surfaces usually have a greater variety of hairs and scales which are diagnostic at the species level.

CYATHEA

Sm. in Mem. Acad. Turin 5: 416 (1793); Holttum in K.B. 36 (3): 463–482 (1981)

Alsophila R.Br., Prodr.: 158 (1810)
Hemitelia R.Br., Prodr.: 158 (1810)

Description as for the family.

Over 600 species distributed throughout the wet tropics, particularly in cool montane forests and the warmer southern temperate region of the world. Poorly represented in dry areas, and few in the north-temperate zone.

Specimens of *Cyathea* in herbaria are usually very poor. Each collection set should include, as a minimum, the stipe and lower lamina, plus the middle and apex of a frond, with label data giving full dimensions.

Descriptions in key and text refer to mature fronds, unless otherwise stated.

1. Fronds pinnate; pinna lobes in some cases separately
   adnate but never truly free (except basal ones), mostly
   crenate 2
   Fronds bipinnate with deeply lobed pinnules (fig. 2, page 10) 6
2. Largest pinnae less than 17 cm long 3
   Largest pinnae more than 20 cm long 4
3. Stipe 2–10 cm long; pinna-midrib beneath with sparse
   hairs; **K** 4, 7; **T** 2, 3, 6, 7 . . . .   1. *C. humilis*
   Stipe more than 20 cm long; pinna-midrib beneath with
   both scales and hairs; **T** 6 . . . .   2. *C. schliebenii*
4. Stipe to 50 cm long, the base covered in long pale scales;
   indusium absent, replaced by ring-like scales; **T** 3
   (naturalised?) . . . .   3. *C. cooperi*
   Stipe 17–30 cm long, the base with dark scales or without
   scales; indusium cup-shaped at maturity 5

5. Stipe 20–30 cm long; lobes of largest pinnae, except a
    few basal pairs, not separately adnate, all widened
    almost symmetrically at their bases; veins once-forked;
    U 2, 4 . . . .   4. *C. camerooniana*
    Stipe 17–20 cm long; lobes of basal half or more of
    largest pinnae separately adnate, distal ones and all
    lobes of upper pinnae decurrent basiscopically; many
    veins twice-forked; **T** 3, 7 . . . .   5. *C. mossambicensis*
6. Pinna-midrib and pinnules beneath with dense
    contorted hairs and some very narrow scales; indusium
    cup-shaped; **U** 2–4; **T** 1–2, 4, 6–8 . . . .   6. *C. dregei*
    Pinna-midrib beneath with broad scales and sometimes
    with a few hairs as well; indusium cup-shaped or absent 7
7. Pinna-midrib beneath with scales and some hairs;
    indusium cup-shaped 8
    Pinna-midrib beneath with only scales, hairs absent;
    indusium absent 9
8. Stipe 25–90 cm long, bearing conical warts or black
    spines 2–4 mm long; largest pinnae 45–75 cm long;
    veins 10–13; widespread . . . .   7. *C. manniana*
    Stipe to 30 cm long, without spines; largest pinnae to
    40 cm long; veins 4–8; **T** 4, 7 . . . .   8. *C. thomsonii*
9. Stipe without spines; pinnae lobes lobulate; **T** 6 . . . .   9. *C. fadenii*
    Stipe with small spines, 1–1.5 mm long; pinnae lobes
    shallowly crenate; **U** 2 . . . .   10. *C. mildbraedii*

1. **Cyathea humilis** *Hieron.* in P.O.A. C: 88 (1895); F.D.-O.A.: 17 (1929); Holttum in
K.B. 36 (3): 476 (1981); K.T.S.L.: 39 (1994); U.K.W.F.: 24 (1994). Types: Tanzania,
Lushoto District: Usambara Mts, Bulwa [Bulua] near Gonja, *Holst* 4264 (B, K!, syn.)
& 4273 (B, syn.)

Caudex absent or up to 2.5 m high and to 4 cm in diameter. Stipe of mature
plants 2–10 cm long, of young plants longer, light brown, not verrucose; basal
scales glossy redbrown, to 10 mm long and 1 mm wide, with dull fragile edges and
setiform apex. Lamina pinnate, 40–180 cm long, texture thin; pinnae 40–50 pairs,
many lower ones gradually decrescent and closely placed, but those of young plants
not decrescent, lowest 1–3 cm long; middle pinnae 2.5–5 cm apart, distinctly
oblique; rachis abaxially glabrescent, adaxially bearing rather sparse light brown
hairs 1 mm long; largest pinnae 12–17 cm long, 2.5–3 cm wide, lobed to 1 mm or
less from costa, basal lobes not free; lobes slightly oblique and slightly falcate, ±
crenate at least distally; costules 4–6 mm apart; veins 7–9 pairs, well spaced, very
slender, forked about one-third above their bases except distal ones. Scales and
hairs: on lower surface of costae rather sparse thick hairs, most abundant distally,
no such hairs on veins but abundant very short appressed hairs present, on upper
surface a few thick hairs distally on costules only. Sori at forks of veins, rusty brown;
indusia thin, quite covering the sorus to maturity or sometimes showing a small
apical aperture, soon breaking and sometimes reduced to an uneven disc round
the bases of the sori.

var. **humilis**

Old fronds persistent; pinnae ± 40 pairs, at oblique angles, the basal ones 1.5–3 cm long;
largest pinnae 12–15 cm long, 2.5–3 cm wide; middle pinnae 3–5 cm apart, at slightly oblique
angle, costules 4.5–6 mm apart. Fig. 1: 2/2A (page 5).

KENYA. Meru District: Nyambeni Hills, 16 km on Maua–Mikinduri road, June 1969, *Faden et al.*
    69/705!; Embu District: Irangi Forest, July 1949, *Schelpe* 2427!; Teita District: Mbololo, May
    1985, *Beentje et al.* 970!
TANZANIA. Kilimanjaro, Weru-weru R. gorge, Dec. 1996, *Hemp* 1375!; Lushoto District: East
    Usambaras, Kwamkoro to Kihuhi, Dec. 1936, *Greenway* 4792!; Morogoro District: Uluguru
    Mts, NW slopes of Bondwa, Aug. 1970, *Pocs* 6226E!
DISTR. **K** 4, 7; **T** 2, 3, 6, 7; Sudan
HAB. Moist forest, often near streams; (?500–)900–2250 m
USES. In Tanzania an infusion of the stem is used as an anthelmintic (fide Greenway)
CONSERVATION NOTES. Fairly widespread but in a declining habitat; data deficient (DD)

SYN. *Alsophila holstii* Hieron. in P.O.A. C: 87 (1895); F.D.-O.A.: 18 (1929); T.T.C.L.: 179 (1949).
        Type: Tanzania, Lushoto District: Usambara Mts, Bulwa [Bulua], *Holst* 4276 (B, holo.,
        K!, iso.)
    Cyathea stuhlmannii Hieron. in E.J. 28: 340 (1900); T.T.C.L.: 180 (1949). Type: Tanzania,
        Morogoro District: Uluguru Mts, Nghweme, *Stuhlmann* 8813 (B, holo., BM, K!, photo.)
    C. ulugurensis Hieron. in E.J. 28: 340 (1900); T.T.C.L.: 180 (1949). Type: Tanzania,
        Morogoro District: Uluguru Mts, *Stuhlmann* 8803 (B, holo.)
    C. opizii Domin., Pteridophyta 263 (1929), *nom. nov., non C. holstii* Hieron. Type as for
        Alsophila holstii Hieron.
    Alsophila stuhlmannii (Hieron.) Tryon in Contr. Gray Herb. 200: 31 (1970)

var. **pycnophylla** *Holttum* in K.B. 36 (3): 478 (1981). Type: Tanzania, Morogoro District:
Nguru South Forest Reserve, *Mabberley & Pocs* 681 (K!, holo., EA!, iso.)

Old fronds not persistent; lamina 115 cm long; pinnae ± 50 pairs, at right angles to rachis,
the basal ones 1–1.5 cm long; largest pinnae 17 cm long, 2.5 cm wide; middle pinnae 2.5 cm
apart, at right angles to costa, separated by narrow sinuses; costules ± 4 mm apart.

TANZANIA. Morogoro District: Nguru South Forest Reserve, E slopes above Kwamanga, Feb.
    1971, *Mabberley & Pocs* 681!; Iringa District: Mwanihana Forest reserve above Sanje village,
    Oct. 1984, *D. Thomas* 3889!
DISTR. **T** 6, 7; not known elsewhere
HAB. Moist forest; ± 1450 m
USES. None recorded from our area
CONSERVATION NOTES. Known from a single specimen; data deficient (DD)

2. **Cyathea schliebenii** *Reimers* in N.B.G.B. 11: 916 (1933); T.T.C.L.: 180 (1949);
Holttum in K.B. 36 (3): 474 (1981). Type: Tanzania, Morogoro District: Uluguru Mts,
Schlieben 3043 (B!, holo.; BM!, iso.)

Trunk 15 cm to 4.5 m tall, massive. Fronds to 1.1 m long; stipe 20 cm or more
long, scaly throughout; basal scales to 12 mm long, 1.5 mm wide near base, dark
reddish brown with almost concolorous fragile margins, apex setiform; scales of
upper part of stipe smaller and less dense. Lamina pinnate (sub-bipinnate), 70–85
cm long; a few pairs of lower pinnae progressively reduced and a little more widely
spaced, lowest 4–7 cm long; largest pinnae to 15 cm long and 3 cm wide; basal pair

FIG. 1. *CYATHEA SCHLIEBENII* — **1**, two middle pinna-lobes of a middle pinna, × 5; **1A**, detail
of veins & costule of 1, × 16; **1B**, one scale, one hair, and one 'intermediate type' from
pinna-midrib, × 13. *CYATHEA HUMILIS* var. *HUMILIS* — **2**, two middle pinna-lobes of a
middle pinna, × 5; **2A**, detail of veins and costule of 2, × 16. *CYATHEA CAMEROONIANA*
var. *UGANDENSIS* – **3**, base of middle pinna-lobe of a middle pinna, × 5. *CYATHEA*
MOSSAMBICENSIS — **4,** base of an middle pinna-lobe of a middle pinna, × 16.
NB. Apex of pinna is to the right in all cases (see arrows); for 1 and 2 the right-hand lobe
is shown with the sporangia removed, only the attachment points are shown. 1 from
Faden 70/656; 2 from *Mabberley* 1239; 3, from *Dawkins* 689; 4, from *Schelpe* 5630. Drawn by
Peter J. Edwards.

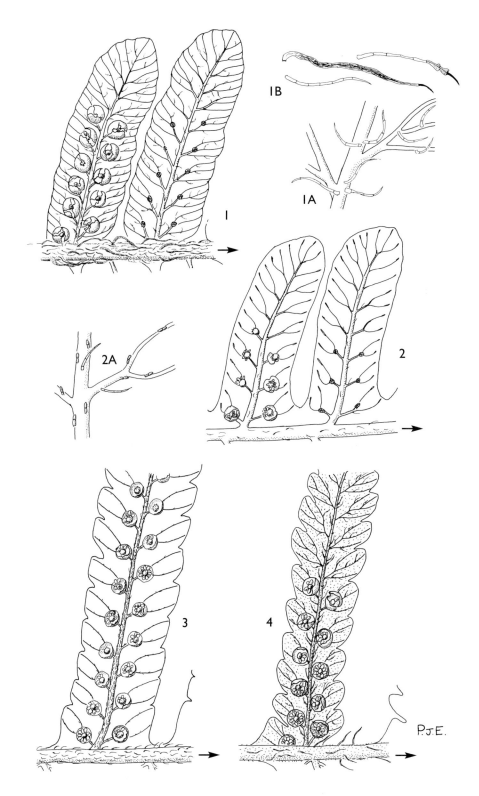

of lobes free; almost all other lobes separately adnate to the costa; costules of lobes 4–5 mm apart; basal (free) lobes of larger pinnae deeply lobulate, other lobes irregularly slightly crenate or almost entire; veins 10–11 pairs, the lower ones twice forked. Scales and hairs on pinnules: lower surface of costa densely covered with narrow brown scales and stiff antrorse and stiff spreading slightly curved hairs; many such hairs present on the lower surface of costules and veins. Sori near costules; indusia almost covering sori to maturity, then almost always breaking; paraphyses short. Fig. 1: 1/1A, B (page 5).

TANZANIA. Mpanda District: Mahali Mts, June 1971, "*M.L.*" 67!; Morogoro District: NW Uluguru
    Mts, Nov. 1932, *Schlieben* 3043! & idem, Bondwa, Sep. 1970, *Faden et al.* 70/656!
DISTR. **T** 4, 6; not known elsewhere
HAB. In roadside banks in areas of wet forest; 1800–2000 m
USES. None recorded from our area
CONSERVATION NOTES. Known from 3 specimens; data deficient (DD), but quite likely very rare

SYN. *Alsophila tanzaniana* Tryon in Contr. Gray Herb. 200: 31 (1970), *nom. nov., non Alsophila*
    schliebenii Reimers in N.B.G.B. 11: 916 (1933)

3. **Cyathea cooperi** (*F.Muell.*) *Domin.*, Pterid.: 262 (1929); Holttum in Blumea 12: 265 (1964); Bostock, Fl. Austral. 48: 203 (1998). Type: Australia, New South Wales, Wollongong, *Woolls* s.n. (MEL, lecto., chosen by Tindale)

Trunk to 10 m high in Australia, in East Africa much smaller. Fronds to 4 m long; stipe to 50 cm long, covered with rounded warts to ± 1 mm long; scales at base mostly white and papery with many small reddish brown marginal and apical setae grading into brown scales, usually deciduous leaving oval scars on the stipe. Lamina pinnate (bipinnatifid); largest pinnae to 65 cm long, up to 26 cm wide; largest pinnules to 12 cm long, 2.5 cm wide, lobed almost to the midrib throughout, a few pairs of basal lobes occasionally free, margins irregularly toothed or rarely deeply lobed. Scales and hairs: pinna rachises, costae and costules below with small white, to reddish setae on margins. Sori 1 to 10, close to midrib or midway between margin and midrib; no true indusium, a ring of white to reddish brown, narrow scales with setiferous margins surrounding the sorus.

TANZANIA. Lushoto District: Amani, Jan. 1950, *Verdcourt* 38!
DISTR. **T** 3; originally from E Australia, but naturalised at Amani and in Western Australia and
    a serious weed in Hawaii
HAB. Roadside banks in forest zone; ± 900 m
USES. None recorded from our area
CONSERVATION NOTES. Widespread; least concern (LC)

SYN. *Alsophila cooperi* F.Muell., Fragm. Phyt. Austr. 5: 117 (1866)

NOTE. The only species present in Africa which does not bear the small 2–3-celled appressed
    hairs typical of most *Cyathea* species. Measurements are partially based on Australian
    specimens which are larger in size than the collection from Amani.

4. **Cyathea camerooniana** *Hook.*, Syn. Fil.: 21 (1865): Alston, Ferns W. Trop. Afr.: 27 (1959); Holttum in K.B. 36 (3): 478 (1981). Type: Cameroon, Cameroon Mts, *Mann* 2059 (K!, holo.)

Trunk to 3 m tall. Frond to 3 m long; stipe 15–30 cm long; scales to 15 mm long, less than 2 mm wide, margins thin, dull. Lamina pinnate, thin-textured, to 2.5 m long; several pairs of lower pinnae gradually reduced and widely spaced, lowest commonly 5–10 cm long; largest pinnae commonly 20–40 cm long, 3–8 cm wide; 1–few pairs basal lobes quite free, the remainder widened at their bases and meeting each other, usually with a distinct costal wing between them, almost at right angles to the costa, crenate at least distally; veins commonly 12–22 pairs, most veins forked once near the costule. Scales and hairs on pinnules: lower surface of costa always bearing stiff, antrorse and stiff spreading hairs, 0.5–1.5 mm long; small scales sometimes present but rarely abundant; scattered, stiff antrorse and spreading stiff hairs present on lower surface of costules. Sori at forks of veins, near but not touching the costules; indusia thin and complete and breaking at maturity or forming cups with definite edges which bear one or more flaccid hairs and scale-like extensions.

SYN. *Alsophila camerooniana* (Hook.) Tryon in Contr. Gray Herb. 200: 30 (1970)

var. **ugandensis** *Holttum* in K.B. 36 (3): 481 (1981). Type: Uganda, Mengo District: Kyagwe, Nakiza Forest, *Dawkins* 689 (K!, holo., BM!, EA!, iso.)

Tufted fern, trunk 0.3–2 m tall, unarmed, densely clothed in scales from leaf-bases. Fronds 0.9–3 m long; stipe 20–30 cm long. Lamina with ± 35 pairs of pinnae; lower pinnae 10-jugate and distant; costulae pubescent beneath with small appressed hairs. Indusia cup-shaped, entire. Fig. 1: 3

UGANDA. Kigezi District: Bwindi National Park, Kayonza, near Ishasha R., Mar. 1995, *Poulsen, Eilu & Nkuutu* 779! & Kinkizi, 6 km SW of Kirima, Sep. 1969, *Lye et al.* 4233!; Mengo District: Gaba, Nov. 1915, *Dummer* 2649!
DISTR. **U** 2, 4; not known elsewhere
HAB. Moist forest, often near streams; 1200–1400 m
USES. None recorded from our area
CONSERVATION NOTES. Restricted to a small area; taxonomic status not resolved; data deficient (DD)

NOTE. Holttum states this taxon resembles *C. humilis* but is distinct in the less reduced and more widely spaced lower pinnae, and in the cup-shaped indusia; all the specimens seen by Holttum and cited in his paper of 1981, except the type, do not have more than 2 separate pinna lobes at the base of the larger pinnae. Their status is therefore ambiguous and will require detailed field studies to understand this.
        Other varieties occur in W Africa.

5. **Cyathea mossambicensis** *Baker* in Ann. Bot. 5: 185 (1891): Schelpe in F.Z. Pterid.: 72, t. 21a (1970) excl. syn. *C. humilis* Hieron.; Holttum in K.B. 36 (3): 475 (1981); Burrows, S. Afr. Ferns: 86, fig. 85 (1990). Type: Mozambique, Namuli, Makua country, *Last* s.n. (K!, holo.)

Caudex to 80(–100) cm tall, to 15–20 cm in diameter. Stipe to 17–20 cm long, dark, minutely tuberculate; basal scales glossy brown with pale fragile margins and dark apical seta. Lamina narrowly elliptic in outline, pinnate to 2-pinnatifid, rarely 3-pinnatifid, to 1.9 m long and 0.8 m wide; rachis light redbrown with scattered pale thick hairs and small scales the abaxial surface; several pairs of lower pinnae gradually decrescent; middle pinnae 6–7 cm apart; largest pinnae (20–)33?cm long, ?3–3.5 cm wide; basal pair of lobes free and widened a little on the basiscopic side at their base; other lobes on basal half of pinnae adnate and decurrent on the basiscopic base, distal lobes connected by their decurrent bases; lobes slightly oblique, lobulate in the middle about one-third towards the costule, less deeply near both base and apex, apex narrowed and obtuse; costules about 7 mm apart; veins to at least 18 pairs in the larger lobes, twice forked except the distal ones. Scales and hairs: lower surface of pinna-midrib near the base bearing some residual small narrow flat light-brown scales, the larger ones with a setiform tip and sometimes lateral setae, distally many spreading pale thick hairs, on the upper surface pale antrorse hairs 1 mm long; lower surface of costules bearing hairs like those of the pinnae-midrib, a few hairs also scattered on the distal veins. Sori in a single row close to each side of the costules of pinnae-lobes, at the first fork of the veins, restricted to the proximal half of each lobe; indusia forming deep thin cups which usually break soon after maturity of the sori, sometimes with a few short slender hairs on the margin; distal paraphyses as long as sporangia and dark at their bases, often abraded. Fig. 1: 4 (page 5).

TANZANIA. Pare District: Chowe Forest reserve, South Pare Mts, 1998, *Hemp* 2260; Lushoto District: Mbalamu [Mbarama], Mar. 1890, *Holst* 2487!; Iringa District: Mwanihana Forest reserve above Sanje village, Oct. 1984, *D.Thomas* 3886!

DISTR. **T** 3, 7; Zimbabwe, Mozambique

HAB. Moist forest; 1400–1850 m

USES. None recorded from our area

CONSERVATION NOTES. Widespread; probably least concern (LC)

SYN. *Cyathea holstii* Hieron., P.O.A. C: 88 (1895); F.D.-O.A.: 17 (1929); T.T.C.L.: 180 (1949). Type: Tanzania, Lushoto District: Usambara Mts, Shagayu Forest near Mbaramu, *Holst* 2487 (B, holo., K!, iso.)
> *Alsophila mossambicensis* (Baker) Tryon in Contr. Gray Herb. 200: 31 (1970); Pic. Serm. in Webbia 27(2): 405 (1972)
> *A. campanulata* Tryon in Contr. Gray Herb. 200: 31 (1970) *nom. nov., non A. holstii* Hieron.; type as for *Cyathea holstii* Hieron.

NOTE. There is no information on the base of the frond.
> Schelpe in F.Z. reports this species from Uganda but there is no evidence for this.

6. **Cyathea dregei** *Kunze* in Linnaea 10: 551 (1836); Hook., Sp. Fil. 1: 23, t.17a (1844); F.D.-O.A.: 17 (1929); T.T.C.L.: 180 (1949); Tardieu, Fl. Madag., Cyath.: 23, t. 6/1 (1951); I.T.U.: 103 (1952); Tardieu in Mem. I.F.A.N. 28: 52 (1953); Alston, Ferns W Trop. Afr.: 27 (1959); Schelpe in F.Z. Pterid.: 74, t.21e (1970); Holttum in K.B. 36 (3): 473 (1981); Burrows, S. Afr. Ferns: 84, fig. 83 (1990). Type: South Africa, between Umzimvubu [Omsamwubo] and Umsicaba [Omsamcaba] Rivers, *Drege* s.n. (LZ†, K!, BM!, L, P, syn.) & Macalisberg, 26°S, *Burke* s.n. (K!, syn.)

Trunk 0.4–5 m tall, massive, to 45 cm diameter or even 90 cm near base; rarely with one or more small branches (forked in *Harley* 9601), many old fronds persisting as a pendulous 'skirt'. Stipe commonly 30–50 cm long; sometimes on the stipe base a pair of very small 1-pinnate pinnae 4–25 cm long; scales reddish brown or purple-brown, to 40 mm long, with very narrow fragile margins which bear an obliquely spreading fringe, helically twisted in the distal part; lamina elliptic or oblong in outline, 3-pinnatifid to 3-pinnate, 0.9–3 m long, 0.6–0.8 m wide, yellow-green above, slightly glaucous beneath; pinnae 12–20 on each side of the rachis, alternate and at an angle, the largest 40–70 cm long; lower pinnae commonly slightly to much reduced, rachis smooth abaxially except near the base; largest pinnules commonly 7 cm long by 1.2 cm wide; basal lobes quite free, the other pinnules lobed almost to the costa; lobes crenate; costules commonly 3–5 mm apart; veins 6–10 pairs, forked near costule except distal ones. Scales and hairs on pinnules: lower surface of costae and costules at first densely covered with slender light brown contorted hair, a few contorted hairs and narrow scales, these also sometimes present on lower surface of veins. Sori at forks of veins, in 2 rows along either side of the costule to 10 per lobe; indusia usually forming intact rather shallow cups, sometimes with small projections on the margin. Fig. 2: 1, 1A (page 10)

UGANDA. Toro District: Butiti, July 1938, *A.S.Thomas* 2326!; Elgon, Namatala R., Oct. 1916, *Snowden* 490!; Masaka District: Sango Bay 6 km N of Kagera R. mouth, Aug. 1951, *Norman* 80!

TANZANIA. Kigoma District: Gombe National Park, Mkenke Valley, Jan. 1964, *Pirozynski* 295!; Iringa District: Mufindi, Idetero House, Mar. 1987, *Lovett* 1725!; Songea District: Halau R. valley 3 km SE of Miyau, Jan. 1956, *Milne-Redhead & Taylor* 8238!

DISTR. **U** 2–4; **T** 1–2, 4, 6–8; widespread in tropical Africa from Sierra Leone to South Africa; Madagascar

HAB. Along streams in grassland, in swamp edges, riverine forest, seepage grassland, pits in ironstone, forest margins, less often in dense forest along streams; may be locally common; 1050–2200 m (–2600 m fide Peter)

USES. None recorded from our area

CONSERVATION NOTES. Widespread; least concern (LC)

SYN. *Cyathea burkei* Hook., Spec. Fil. 1: 23, t.17b (1844). Type: South Africa, Macalisberg, *Burke* s.n. (K!, holo.)

    C. angolensis Hook., Syn. Fil. 1: 22 (1865). Type: Angola, Benguela, Huila, *Welwitsch* 83 & 186 (K!, BM!, syn.)

    Alsophila dregei (Kunze) Tryon in Contr. Gray Herb. 200: 30 (1970)

NOTE. *Stolz* 907 from Tanzania, Rungwe District: Kyimbila seems a natural hybrid *C. thomsonii* Bak. × *C. dregei* Kunze.

7. **Cyathea manniana** *Hook.* in Hook. & Bak., Syn. Fil.: 21 (1865); F.D.-O.A.: 17 (1929); Tardieu in Mem. I.F.A.N. 28: 5, t. 6/3–5 (1953); Alston, Ferns W Trop. Afr.: 27, t. 8 (1959); Schelpe in F.Z. Pterid.: 72, t. 24b (1970), excl. *C. deckenii* Kuhn.; Holttum in K.B. 36 (3): 472 (1981); Burrows, S. Afr. Ferns: 87, fig. 86 (1990); K.T.S.L.: 39, map (1994); U.K.W.F.: 24 (1994). Type: Equatorial Guinea, Bioko [Fernando Po], *Mann* 363 (K!, syn.) & Cameroon Mts, *Mann* 1392 (K!, syn.)

Trunk 0.3–9 m tall, 10–15 cm in diameter, dark brown to almost black, throughout with long-decurrent appressed spiny stipe bases, in older trees the lower stem eventually smooth. Fronds 5–10 in number, to 4 m long; stipe brown, 25–90 cm long, bearing conical warts or black spines 2–4 mm long; scales medium to dark reddish brown, to 20 mm long and 6 mm wide near the base, margins fragile with irregularly projecting oblique thin-walled cells. Lamina dark green, 2-pinnate, 3-pinnatifid or 3-pinnate, 0.9–3 m long, 0.9–1.3 m wide; basal pinnae not or little reduced, long stalked; rachis spiny towards its base, with (12–)35–40 falcate pinnae on each side; largest pinnae 45–75 cm long; largest pinnules 6–13 cm long and 1–2.2 cm wide, ± glaucous beneath, basal pair of lobes free, remaining pinnules lobed almost to the costa, the lobes crenate; costules 2.5–4 mm apart; veins 10–13 pairs, lower and middle ones of large pinnules twice forked, distal ones forked or simple. Scales and hairs on pinnules: lower surface of costae at first covered by broad flat light- to mid-brown scales of all sizes, the larger ones with a dark median patch, all with fringed edges, the scales deciduous on old fronds, grading to a usually small number of short contorted hairs of variable length and narrow hair-like scales; lower surface of costules scaly and hairy as costae. Sori close to costules; indusia light brown, broadly cup-shaped with a ± permanent smooth edge or rather one-sided due to splitting; paraphyses short and slender. Fig. 2: 2, 2A, B (page 10)

UGANDA. Kigezi District: Ishasha Gorge 6 km SW of Kirima, Sep. 1969, *Faden et al.* 69/1205!; Toro District: Bwamba Pass, Nov. 1935, *A.S.Thomas* 1451!; Mbale District: Butandiga to Bulago, Dec. 1938, *A.S.Thomas* 2546!

KENYA. Kiambu District: Gatamayu Forest, Mar. 1964, *Verdcourt* 3992!; Kiambu District: Uplands Forest Station NE of Limuru, 1970, *Faden* 70/387!; Teita District: Mbololo, Aug. 1938, *Joanna* in *CM* 9027!

TANZANIA. Tanga District: Kwamkoro Forest Reserve, Jan. 1961, *Semsei* 3176!; Mpanda District: Musenabantu, Aug. 1959, *Harley* 9364!; Songea District: Liwiri-Kiteza Forest Reserve, Nov. 1951, *Eggeling* 6376!

DISTR. U 2–4; K 1–7; T 2–8; widespread in tropical Africa from Liberia to Kenya and south to Mozambique and Zimbabwe

HAB. Gregarious, in forest along streams or adjacent to springs, often forming a secondary canopy in high rainfall areas, occasionally going up into the tree heath zone; 850–2700 m

USES. In Tanzania used for building poles (fide Greenway, Semsei)

CONSERVATION NOTES. Widespread; least concern (LC)

SYN. *Cyathea deckenii* Kuhn in v.d. Decken, Reisen, Bot. 3, 3: 57 (1879); T.T.C.L.: 179 (1949); I.T.U.: 103, photo. 15 (1952); Holttum in K.B. 36 (3): 471 (1981). Type: Tanzania, Mt Kilimanjaro, Dschogge region (?= Chagga?), *v. d. Decken & Karsten* 72 (B, holo.; K, photo!)

    C. usambarensis Hieron., P.O.A. C: 88 (1895); F.D.-O.A.: 17 (1930). Type: Tanzania, Lushoto District: Usambara Mts, Shagayu Forest at Mbaramu, *Holst* 2498 (B!, holo., K!, iso.)

    C. laurentiorum Christ in De Wild., Miss. E. Laurent 1: 14 (1905). Type: Congo (Kinshasa), Butala swamp, *Laurent* s.n. (BR, holo.)

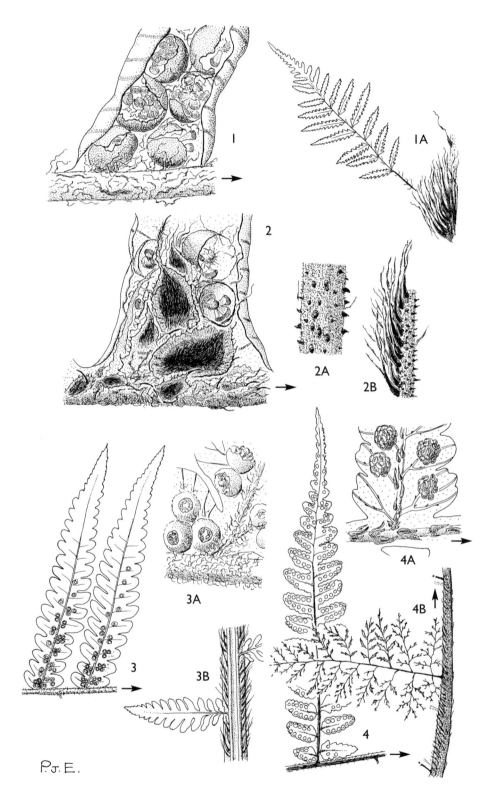

PⱼE.

C. sellae Pirotta in Ann. Bot. Roma 7: 173 (1908). Type: Uganda, Ruwenzori, Mubuku Valley, Nakitava to Bihunga, *Duc. Abruzzi Exped.* s.n. (RO, holo.)

Alsophila manniana (Hook.) Tryon in Contr. Gray Herb. 200: 30 (1970)

A. deckenii (Kuhn) Tryon in Contr. Gray Herb. 200: 30 (1970)

Cyathea manniana sensu Schelpe, F.Z. Pterid.: 72 (1970) pro parte

NOTE. *C. deckenii* has been separated from *C. manniana* by many authors. The very reduced pinnae near the stipe base and the obviously one-sided (hemitelioid) indusia, were used to distinguish it from that species. Holttum (1981) recognised 4 groups and made an important distinction between species with hemitelioid sori and sori with a complete cup or disc. The type collection of *C. deckenii* Kuhn consists of three detached parts: one complete small pinna and two incomplete pinnae, both without sori.

Pichi Sermolli (1956) in comparing *deckenii* with *manniana*, cited only the type and *Alluaud* 349 as representing *deckenii*. Both are from Kilimanjaro. He considered *deckenii* a distinct and probably highly localized species, and that it "was misunderstood by authors and nearly all the specimens from Tropical East Africa referred to this species are wrongly identified", and were in fact *manniana*. The character states he used seem weak to me, as such can be seen to vary within a single frond of the better collections of *manniana* from E Africa (mostly made since 1956), and indeed the rest of Africa in which it occurs. I consider the poor type material could very well have come from near the apex of a frond, where costule distance and other parameters he uses are smaller.

Holttum (1981) also upheld *deckenii* as a good species, without reference to the Pichi Sermolli paper. He cited five additional collections which he considered *deckenii*, from three widely separated locations in Tanzania, Congo Republic and Mozambique. I have examined five of these, including *Allaud* 349, and find that despite Holttum's emphasis on the presence of a hemitelioid (one-sided) indusium in *deckenii* (which of course could not be ascertained from the rudimentary and sterile type collection), all of these collections have entire indusial cups, albeit most of them variously split in one or two places and therefore can **appear** to be one-sided. In addition, the folder at K containing *Allaud* 349 (2 sheets) has an annotation on it, in Holttum's hand, as follows "these specimens are not closely distinct from *manniana*". This doubt is not hinted at in his 1981 paper. On checking this material through the Beck microscope and illuminator Holttum was using at Kew to examine African *Cyathea* in 1980–81, the images possible are indeed conclusive on this point. In addition, in his description of *deckenii* (enlarged to take in the seven collections he cites) he mentions the presence of reduced basal pinnae, presumably alluding to *Faden* 70/314, which is the only specimen of the six he saw to have a stipe base. Incidentally what appeared to him to be the base of a reduce pinna on the K set of this number is in fact an aborted pinna. Those two character states were used by Holttum (see p. 464 of his 1981 paper) to position *deckenii* in his 'group 2', along with *C. capensis*, which demonstrably has a one-sided indusium. Hence the position of these two names together in his key. *C. manniana* he placed in his 'group 3', with *dregei* and *thomsonii*, all of which he stated as *not* having reduced pinnae. My study of African collections in BM, EA, K, WAG, in particular six very good collections from East Africa (some of which were collected after 1981 or were otherwise apparently not seen by Holttum) show clearly that *manniana* can bear one or more (usually two) very reduced

FIG. 2. *CYATHEA DREGEI* — **1**, base of a middle lobe of a middle pinnule of a middle pinna, × 16; **1A**, one of the two reduced basal pinnae, from 10 cm above the base of the stipe, × ²/₃. *CYATHEA MANNIANA* — **2**, base of an middle lobe of a middle pinnule of a middle pinna, × 16; **2A**, base of a large stipe (scales mostly lost) from 10 cm above the base of the stipe, × ²/₃; **2B**, base of a small stipe, from 10 cm above the base of the stipe, × ²/₃. *CYATHEA THOMSONII* — **3**, two middle pinnules of a middle pinna (veins not shown), × 1; **3A**, detail of base of an lobe from 3, × 6; **3B**, the two very reduced basal pinnae, adaxial view, × ²/₃; *CYATHEA FADENII* — **4**, one middle pinnule of a middle pinna, × 1; **4A**, detail of base of an lobe from 4, × 6; **4B**, one of the 3 pairs of basal reduced pinnae ('aphlebia'), lateral view of stipe, the bases of the other 5 are shown, × ²/₃ (note that the two pinnae above these are reduced normal pinnae).

NB. Apex of the pinnule is to the right for 1, 2, 3A, & 4A; apex of the pinna is to the right in 3 & 4 (see arrows). 1 from *Harley* 9601, 1A from *Lovett* 1725; 2 & 2A from *Faden* 70/387, 2B from *Mabberley & McCall* 19; 3 from *Newbould & Jefford* 1932; 4 & 4A from *Schippers* T 17258, 4B from *Pocs & Lundquist* 6476. Drawn by Peter J. Edwards.

pinnae, or aborted such, near the stipe base. The aborted pinnae may be very small and inconspicuous. *Hepper & Field* 5186 from the West Usambaras is particularly revealing, with its long stipe and a single reduced pinna 5 cm long by 1.5 cm wide, plus a tiny aborted pinna. It also shows sori with complete indusial cups directly abutting onto sori with cups tilted and/or split in such a way as to appear hemitelioid.

Interestingly, Schelpe (1970) placed *deckenii* in synonymy under *manniana*, without comment. In Zimbabwe *Asplenium hypomelas* seems restricted to the stems of *C. manniana*.

8. **Cyathea thomsonii** *Baker* in J.B. 19: 180 (1881); Schelpe in F.Z. Pterid.: 72, t. 21c (1970); Holttum in K.B. 36 (3): 473 (1981); Burrows, S. Afr. Ferns: 86, fig. 84 (1990). Type: Tanzania, Njombe/Rungwe District: lower plateau N of Lake Nyassa, *Thomson* s.n. (K!, holo.)

Trunk erect, to 2.5 m tall and to 15 cm in diameter. Fronds to 2 m long and 80 cm wide; stipe to 30 cm long; basal scales to 20 mm long, dark reddish brown, with narrow thin fringed edges. Lamina elliptic in outline, 2-pinnate to 3-pinnatifid with varying degrees of dissection, to 2 m long, a few pairs of lower pinnae gradually reduced and more widely spaced, lowest 5–12 cm long; largest pinnae commonly 40 cm long, pinnules 10–14 mm apart; largest pinnules commonly to 6 cm long by 1 cm wide, basal 1–4 pairs with stalks less than 1 mm long, the others sessile, lobed to 1 mm from the costa, often some pinnules gradually contracted towards the base, but in all cases the basal pair of lobes conspicuously longer; costules 2.5–3 mm apart: veins commonly 4–5 pairs except in basal lobes, on largest fronds 7–8 pairs, lower ones forked near the costa. Scales and hairs on pinnules: on lower surface of costae many light brown hairs and narrow brown scales, the scales grading distally to hairs; costules and veins bearing pale less contorted to stiff, spreading and almost antrorse hairs on the lower surface. Sori near costules; up to 5 in basal lobes, usually one in each distal lobe, indusia at first forming deep cups with an uneven rim, soon breaking irregularly. Fig. 2: 3, 3A, B (page 10).

TANZANIA. Mpanda District: Mahali Mts, Sisaga, Aug. 1958, *Newbould & Jefford* 1932!; Iringa District: N part of Gologolo Mts, Sep. 1970, *Thulin & Mhoro* 960!; Rungwe District: Rungwe, Sep. 1912, *Stolz* 884!
DISTR. **T** 4, 7; Congo (Kinshasa), Angola, Zambia, Malawi, Mozambique, Zimbabwe
HAB. Along streams in moist forest; 1200–2250 m
USES. None recorded from our area
CONSERVATION NOTES. Widespread; least concern (LC)

SYN. *Cyathea zambesiaca* Baker in Ann. Bot. 8: 121 (1894). Type: Malawi, *Buchanan* s.n. (K!, holo.)
    Alsophila thomsonii (Bak.) Tryon in Contr. Gray Herb. 200: 31 (1970); Pic. Serm. in Webbia 27: 403 (1972)

NOTE. A variable species sometimes confused with *C. dregei* with which it is reported to hybridize freely [Burrows 86 (1989)]. It can be distinguished from *C. dregei* by its narrower and shorter caudex; the stipe scales on the sheath and the fusion of many pinnules to the pinna-midrib. Holttum (1981) in his discussion of the four African groups he recognised, stated *C. thomsonii* looks like a hybrid or a series of hybrids between *C. dregei* and one or more species of his fourth group (the simply pinnate species). *Stolz* 907 from Tanzania, Rungwe District: Kyimbila seems a natural hybrid *C. thomsonii* Bak. × *C. dregei* Kunze.

9. **Cyathea fadenii** *Holttum* in K.B. 36 (3): 470 (1981). Type: Tanzania, Morogoro District: Uluguru Mts, *Schlieben* 2976 (B!, holo., BM!, iso.)

Trunk 1–3 m tall, 3.5–5 cm diameter, bearing persistent stipe bases, but these eventually deciduous. Fronds 1.5–2.5 m long; stipes 12–33 cm long, bearing near base of stipe 2 to many small 3–4-pinnate, flaccid and skeletonised or partly skeletonised reduced pinnae, 5–17 cm long, occasionally above these 1 or 2 reduced but more normal pinnae with obvious lamina. Lamina bipinnate, 1.6–2.2 m long, ± 0.6 m wide; largest pinnae 24–48 cm long; largest pinnules 5–9 cm long by 1–2.2 cm wide; basal 1 to 2 pairs of lobes free, sometimes minutely stalked, the remaining costules also 4–5 cm apart, but connected by a narrow green wing; basal lobes lobulate half way to the costule, the remainder gradually less deeply cut to crenate; veins 5–7 pairs, forked once or twice. Scales on lower surface of pinnules; on lower surface of costae near base reddish brown, narrow, to 2 mm long, grading into smaller, more scattered, brown, bullate-lanceolate scales, costules commonly with bullate scales, veins with scattered appressed hairs. Sori at the forks of veins, exindusiate; paraphyses slender, as long as the sporangia. Fig. 2: 4, 4A, B (page 10).

TANZANIA. Morogoro District: Uluguru, SE ridge of Bondwa, Sep. 1970, *Faden et al.* 70/636! & Lupanga Ridge, Feb. 1970, *Pocs & Harris* 6131d!; Kilosa District: Ukaguru [Itumba], Mar. 1905, *North Wood* s.n.
DISTR. **T** 6; endemic to the Uluguru and Ukaguru
HAB. Montane forest, high altitude mist forest with *Melchiora schliebenii, Allanblackia sacleuxii* and *Allanblackia uluguruensis*, extending up into the subalpine heath; 1700–2100 m
USES. None recorded from our area
CONSERVATION NOTES. Area of occurrence restricted to less than 20,000 km², known from 7 specimens; vulnerable (VU-B1ab(iii))

SYN. *Alsophila schliebenii* Reimers in N.B.G.B. 11: 918 (1933); T.T.C.L.: 179 (1949), *non Cyathea schliebenii* Reimers in N.B.G.B. 11: 918 (1933)

10. **Cyathea mildbraedii** (*Brause*) *Domin.*, Pteridophyta: 263 (1929); Holttum in K.B. 36 (3): 469 (1981). Type: Congo (Kinshasa) Uganda, Ruwenzori, Butahu [Butagu] Valley, *Mildbraed* 2545 (B, holo., K!, photo)

Caudex to 4 m tall. Stipe 56 cm long, very dark, bearing copious blunt spines 1–1.5 mm long on the lower half; scales very dark, glossy with dull fragile margins, to 10 mm long; one pair of reduced pinnae near base of stipe (represented by broken bases only, in the specimens studied). Lamina incompletely known; basal pinnae 29 cm long, next pair 32 cm long, others to 40 cm long; rachis abaxially dark, glabrous, minutely verrucose, hairs on adaxial surface brown, thin, somewhat contorted; near base of pinna-rachis some scales 3–4 mm long like those on the stipe, distally some entirely thin scales; pinnules to 8 cm long, 1.8 cm wide, shortly acuminate, 2 cm apart; basal pair of lobes free, then one pair adnate, rest of pinnule lobed almost to the costa; lobes slightly oblique and slightly falcate, the larger ones shallowly crenate throughout; costules 4–4.5 mm apart; veins to 8 pairs, mostly forked below the middle, slender, slightly prominent on the lower surface, not on the upper. Scales and hairs: on lower surface of costae brown flat elongate scales near base grading distally to brown bullate scales, no hairs; on costules brown bullate hair-pointed scales; on upper surface of costae contorted hairs as on pinna-rachis, costules hairless except near base of pinnules. Sori at forks of veins, exindusiate; receptacle short; paraphyses short.

UGANDA. *Osmaston* 3345 is said to have been collected in Uganda; I have not seen the specimen.
DISTR. **U** 2; Congo (Kinshasa), endemic to Ruwenzori Mts
HAB. Heath forest on ridge crest; ± 3300 m
USES. None recorded from our area
CONSERVATION NOTES. Known from just 2 specimens, both from the Ruwenzori; possibly Endangered (EN-D)

SYN. *Alsophila mildbraedii* Brause in Z.A.E. 1: 2, t. 1 (1910)

Cyathea capensis (*L.f.*) *Sm.* in Mem. Acad. Turin 5: 417 (1793): Schelpe in F.Z. Pterid.: 74, t. 21d (1970); Holttum in K.B. 36 (3): 470 (1981); Burrows, S. Afr. Ferns: 88, fig. 87 (1990). Type: South Africa, Cape of Good Hope, *Sparrmann* s.n. (LINN 1251/61, holo.)

This species has been reported from S Tanzania in Flora Zambesiaca, by Holttum in Kew Bulletin, and by Burrows in his Southern African Ferns; the editor believes all this is due because of the mention in Peter's Flora von Deutsch-Ostafrika: 18 (1929). However, Peter only cites his own collection from Knysna (South Africa). This species has never been found north of Mt Mlanje.

INDEX TO CYATHEACEAE

PLANTS PEOPLE
POSSIBILITIES

First published in 2005 by
Royal Botanic Gardens, Kew
Richmond, Surrey, TW9 3AB, UK
www.kew.org

ISBN 1 84246 114 1

Design by Media Resources, typesetting and page layout by Margaret Newman,
Information Services Department,
Royal Botanic Gardens, Kew.

Printed by Cromwell Press Ltd.

For information or to purchase all Kew titles please visit
www.kewbooks.com or email publishing@kew.org